Sound

Copyright © by Harcourt, Inc.

All rights reserved. No part of this publication may be reproduced or transmitted in any form or by any means, electronic or mechanical, including photocopy, recording, or any information storage and retrieval system, without permission in writing from the publisher.

Requests for permission to make copies of any part of the work should be addressed to School Permissions and Copyrights, Harcourt, Inc., 6277 Sea Harbor Drive, Orlando, FL 32887-6777. Fax: 407-345-2418.

HARCOURT and the Harcourt Logo are registered trademarks of Harcourt, Inc., registered in the United States of America and/or other jurisdictions.

Printed in Mexico

ISBN 978-0-15-362048-5
ISBN 0-15-362048-X

6 7 8 9 10   0908   16 15 14 13 12

4500358761

Harcourt
SCHOOL PUBLISHERS

Visit *The Learning Site!*
www.harcourtschool.com

Lesson 1

VOCABULARY
vibration
pitch
intensity

What Is Sound?

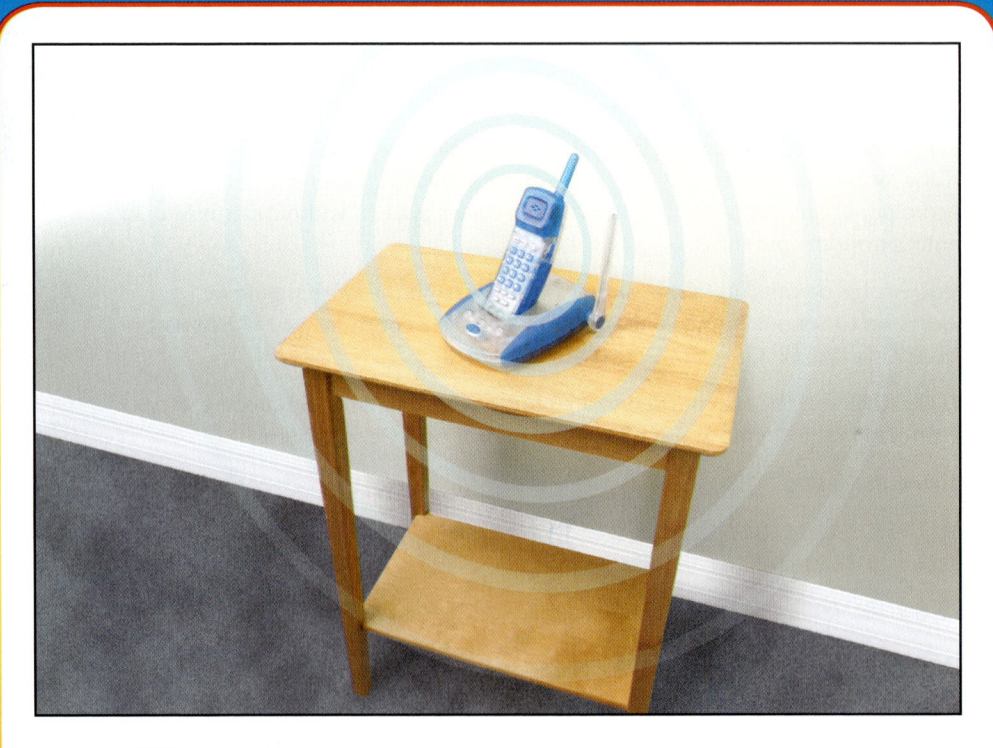

A **vibration** is a quick back-and-forth motion. Vibrations that cause sound are called sound waves.

Lower Pitch ⟵⟶ **Higher Pitch**

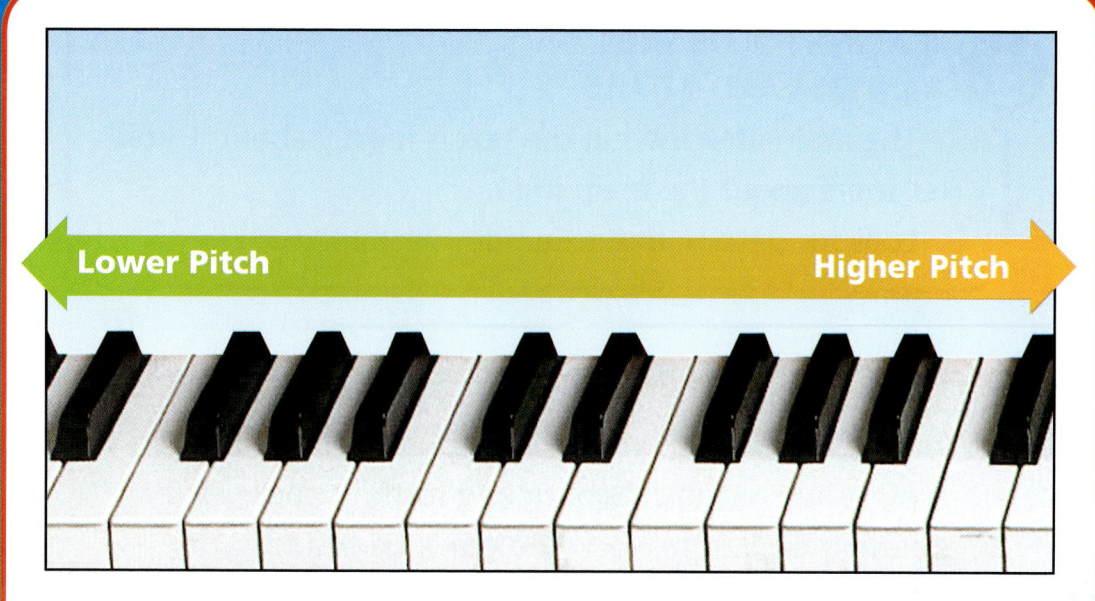

Pitch is how high or low a sound is. How fast or slow an object vibrates determines pitch.

Intensity is the measure of how loud or soft a sound is. Highway traffic has a greater sound intensity than leaves blowing in the wind.

READING FOCUS SKILL
MAIN IDEA AND DETAILS

The main idea is what the text is mostly about. Details tell more about the main idea.

Look for details about sound.

Sources of Sound

What happens when you pluck a guitar string? It vibrates, or moves quickly back and forth. It causes the air around it to vibrate, too. The quick back-and-forth motion is a **vibration**. Vibrations that make sound are called sound waves. You hear sound when sound-waves reach your ear.

What produces sound?

Vibrations from a speaker in the phone move through the air to your ear.

Pitch

Pitch is how high or low a sound is. Things that vibrate quickly make high sounds. Those that vibrate slowly make low sounds.

Small objects usually vibrate quickly. They make sound waves that are close together. Large objects usually vibrate more slowly. They make sound waves that are spread out. Close-together waves sound higher than the spread-out ones.

Focus Skill: Tell what kind of waves makes low-pitched sounds.

◀ Long, thick strings of a bass make a low sound.

Tiny vocal cords of a bird make a high sound. ▶

Lower Pitch ⬅ ➡ Higher Pitch

The strings on the left side of the piano vibrate more slowly.

Intensity

Intensity is the measure of how loud or soft a sound is. Sound waves with more energy make louder sounds. A jet engine makes sound waves with great energy. The sound waves make a very loud sound. A pin dropping makes sound waves with much less energy. These sound waves make a much softer sound.

What kind of waves makes loud sounds?

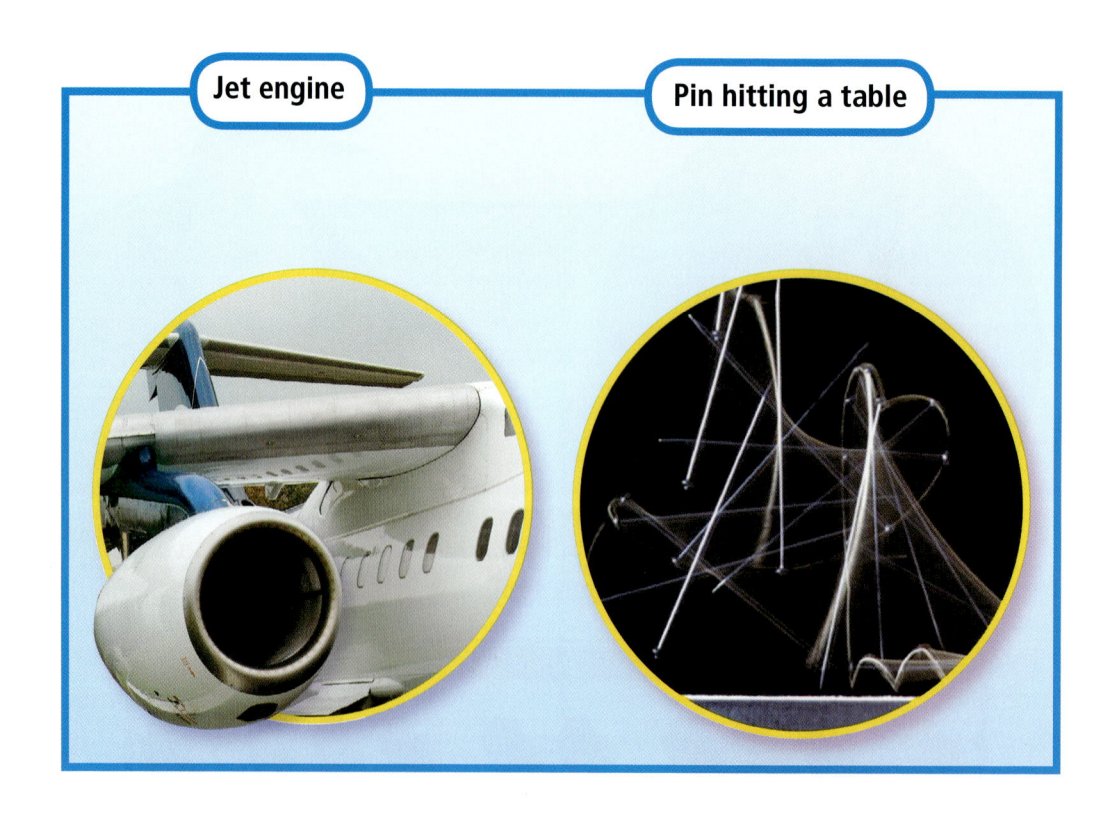

Jet engine

Pin hitting a table

Measuring Sound Intensity

Sound intensity is measured in decibels (dB). Any increase of 10 decibels means a sound is ten times louder.

What is a decibel? How is it used?

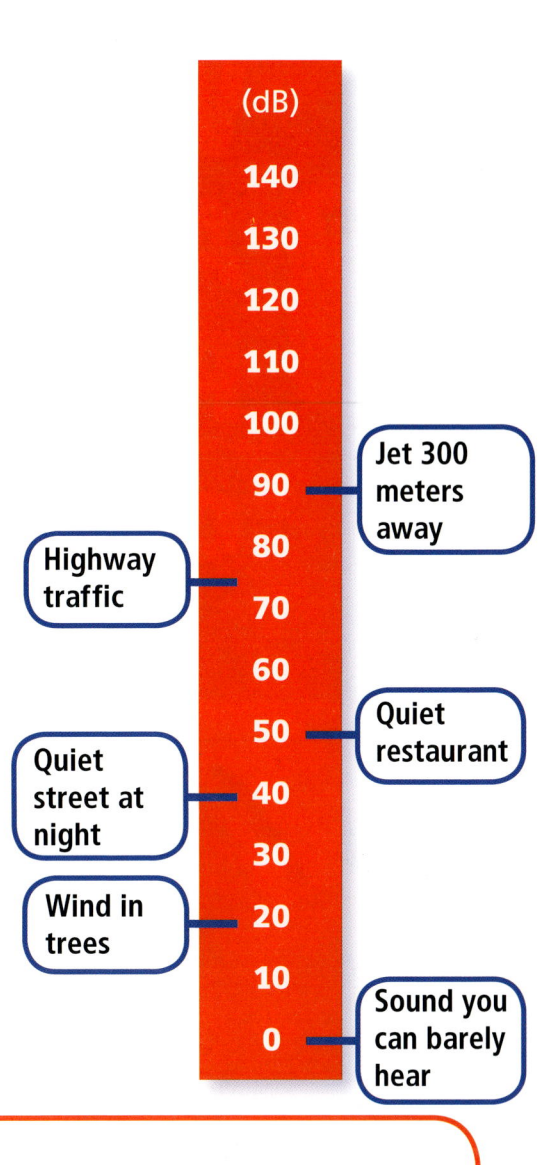

(dB)
- 140
- 130
- 120
- 110
- 100
- 90 — Jet 300 meters away
- 80
- 70 — Highway traffic
- 60
- 50 — Quiet restaurant
- 40 — Quiet street at night
- 30
- 20 — Wind in trees
- 10
- 0 — Sound you can barely hear

Review

Complete this main idea statement.

1. Sound is produced by _____.

Complete these detail statements.

2. Small objects vibrate _____.

3. Sound waves with more energy make _____ sounds.

4. Most sounds are measured in _____.

7

Lesson 2
What Are the Properties of Waves?

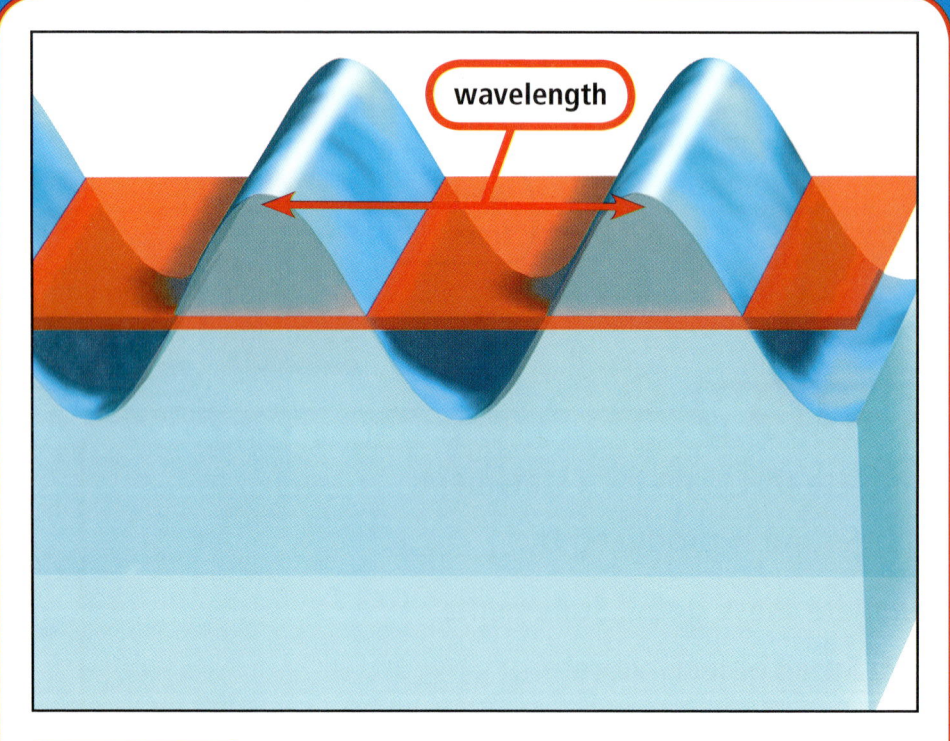

VOCABULARY
wavelength
frequency
amplitude

wavelength

Wavelength is the distance between a point on one wave and the identical point on the next wave.

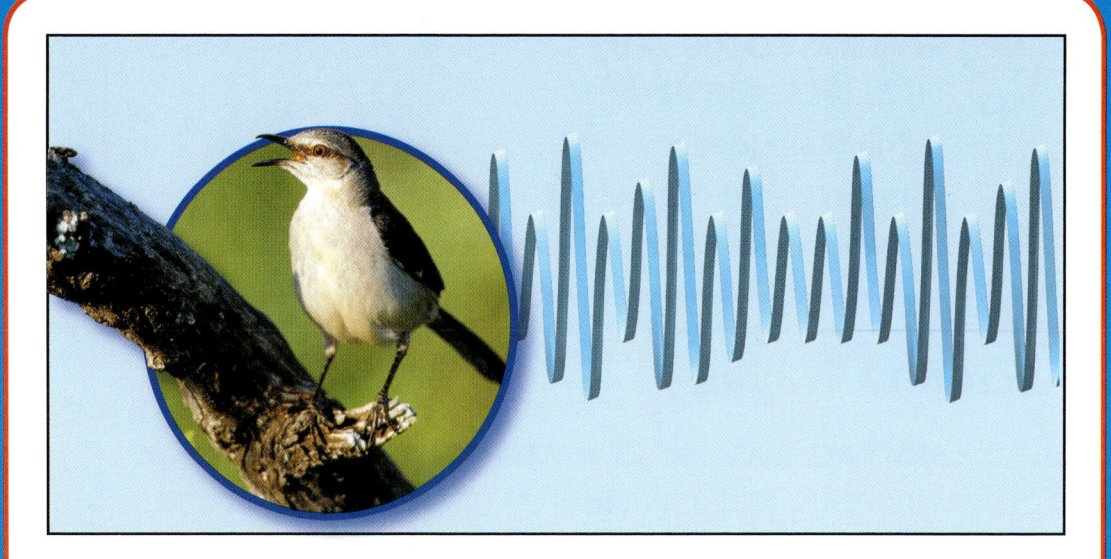

Frequency is a measure of the number of waves that pass in a second. Sound waves from a bird have a high frequency.

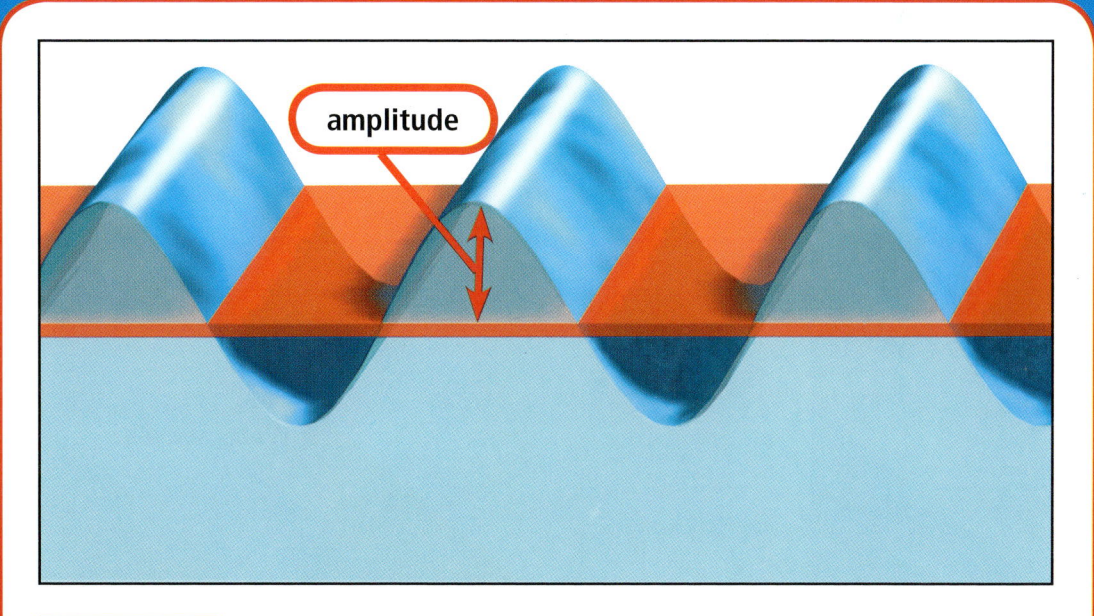

amplitude

Amplitude is a measure of the amount of energy in a wave. Sound waves with large amplitudes are louder.

READING FOCUS SKILL
MAIN IDEA AND DETAILS

The main idea is what the text is mostly about. Details tell more about the main idea.

Look for details about sound waves.

Sound Waves

If you have ever watched ocean waves, you know that they move up and down as they move forward. Sound moves in waves, too, but they do not move up and down.

Have you ever played with a spring toy? If you hold one end and move it in and out, you see a section of bunched-up coils moving down to the end and back. That's what sound waves are like.

◀ Water waves

The three properties of waves are wavelength, frequency, and amplitude. **Wavelength** is the distance between a point on one wave and the same point on the next wave. **Frequency** is a measure of the number of waves that pass in a second. **Amplitude** is a measure of the amount of energy in a wave.

What are three properties of waves?

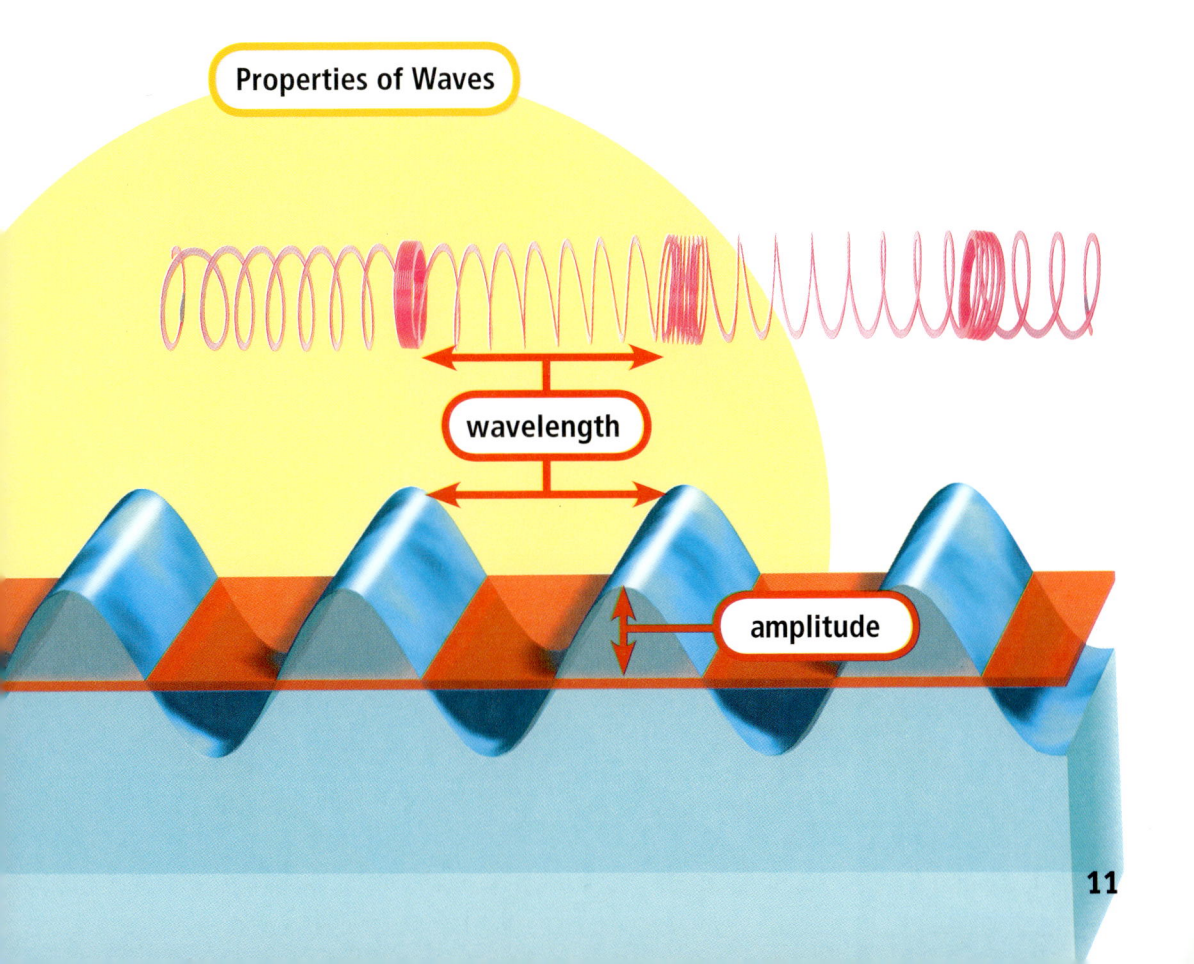

Properties of Waves

wavelength

amplitude

Frequency and Pitch

If you could see sound waves, you could count how many go past in a second. That would tell you their frequency. The more waves that go past in one second, the higher the frequency.

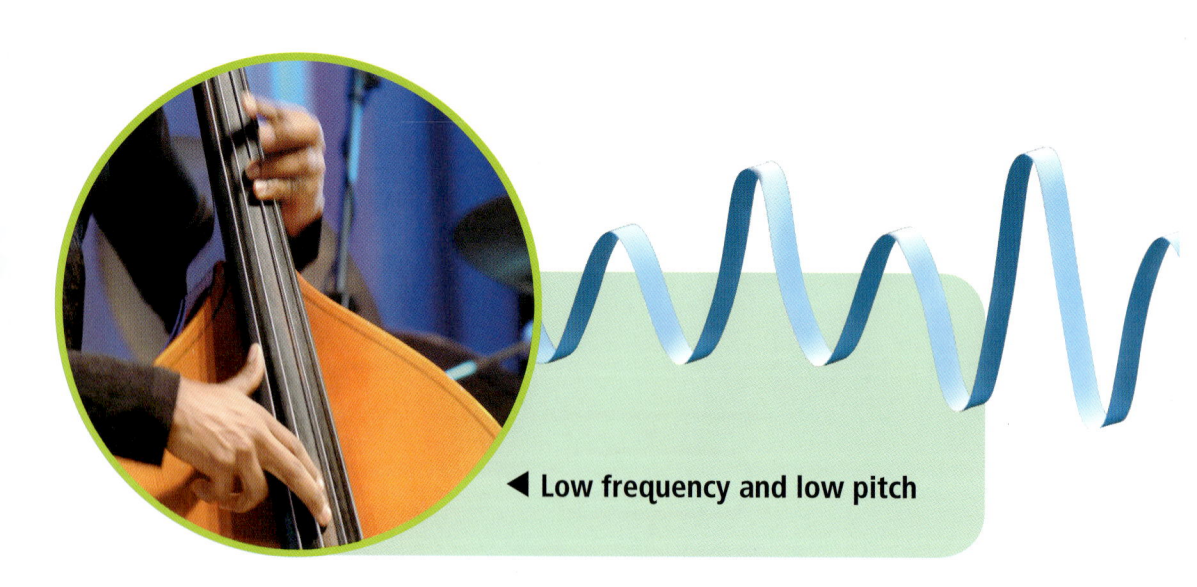

◀ **Low frequency and low pitch**

If sound wave peaks go past quickly, they are close together. That means the sound source is vibrating quickly. Things that vibrate quickly make a high pitch. The higher the frequency of the waves, the higher the pitch of the sound.

Tell how frequency and pitch are related.

◀ **High frequency and high pitch**

Amplitude and Loudness

The volume of a sound is called intensity. Intensity and amplitude are related.

Sound waves with low intensity are soft. They have small amplitudes.

Sound waves with high intensity are the opposite. They are loud. They have large amplitudes.

Focus Skill **How does the amplitude of a sound relate to its intensity?**

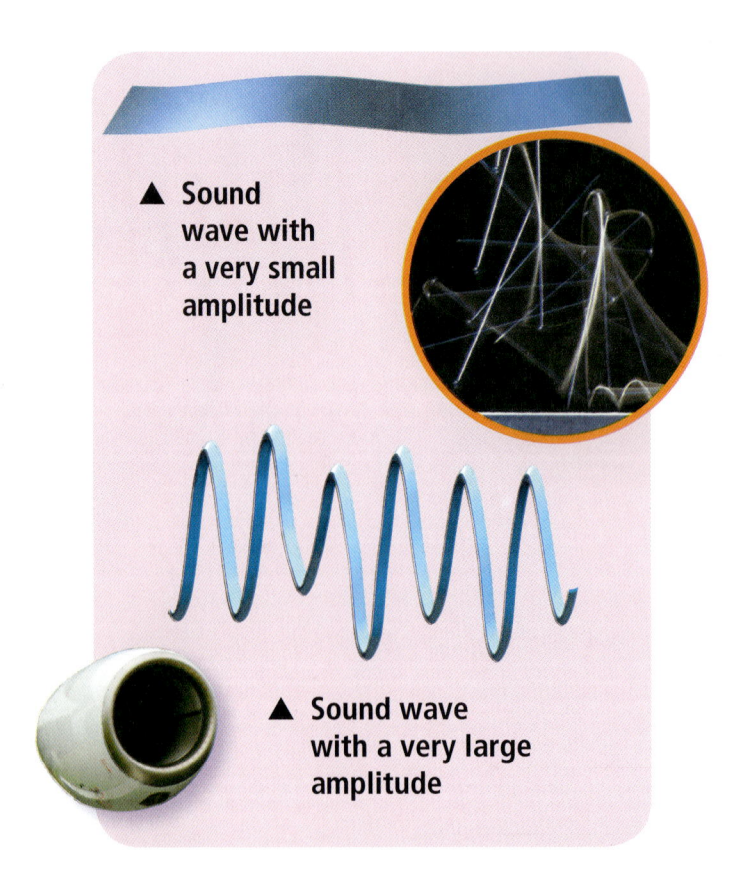

▲ Sound wave with a very small amplitude

▲ Sound wave with a very large amplitude

Measuring Sound Waves

You can measure the properties of sound waves. An instrument can show sound waves. It has a microphone. The microphone picks up sound waves. The instrument changes the sound waves into lines on a screen.

Tell what machine you can use to measure the properties of sound waves.

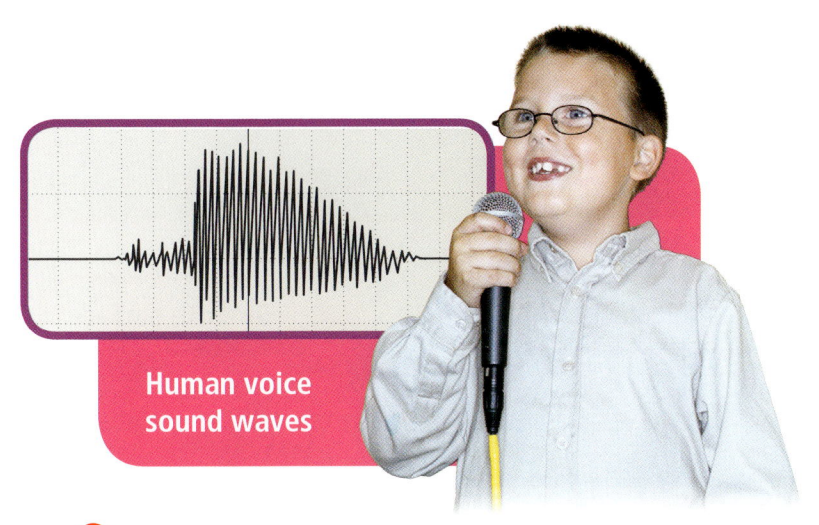

Human voice sound waves

Review

Complete this main idea statement.

1. Three properties of a wave are wavelength, frequency, and _____.

Complete these detail sentences.

2. Sound waves that have a high frequency have a high _____.

3. Sound waves with small amplitudes are _____.

4. Sound waves with large amplitudes are _____.

Lesson 3

How Do Sound Waves Travel?

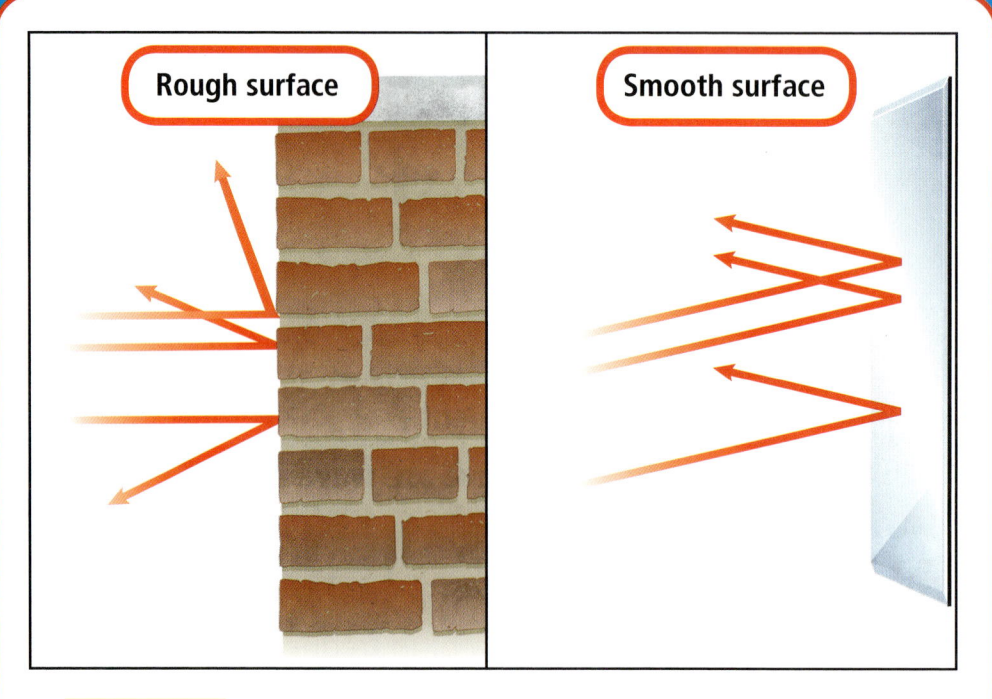

VOCABULARY
reflection
absorption
transmission

Rough surface | **Smooth surface**

A **reflection** is the bouncing of light, sound, or heat off an object. A flat smooth surface keeps the pattern of sound waves. A rough surface does not.

Absorption is the taking in of sound by an object. When sound waves are absorbed, they are stopped. This room absorbs any sounds made in it.

Transmission is the passing of sound through a material. Water passes sound. This is how whales hear each other.

READING FOCUS SKILL
COMPARE AND CONTRAST

When you **compare and contrast**, you tell how things are alike and different.

Look for ways to **compare and contrast** how sound waves interact with different objects.

Hearing Sounds

When a rocket is launched into space, it makes a huge roar. Sound waves go out in all directions. They travel through matter to your ears.

▲ Sound waves travel through air and the pillow to reach the girl's ears.

When you hear sound, your outer ear directs sound waves into your eardrum. The sound waves make your eardrum vibrate. The vibrations move along three tiny bones to the inner ear. In an organ called the cochlea, cells change the vibrations to signals. The signals travel along a nerve to your brain. Your brain recognizes the signals as sound.

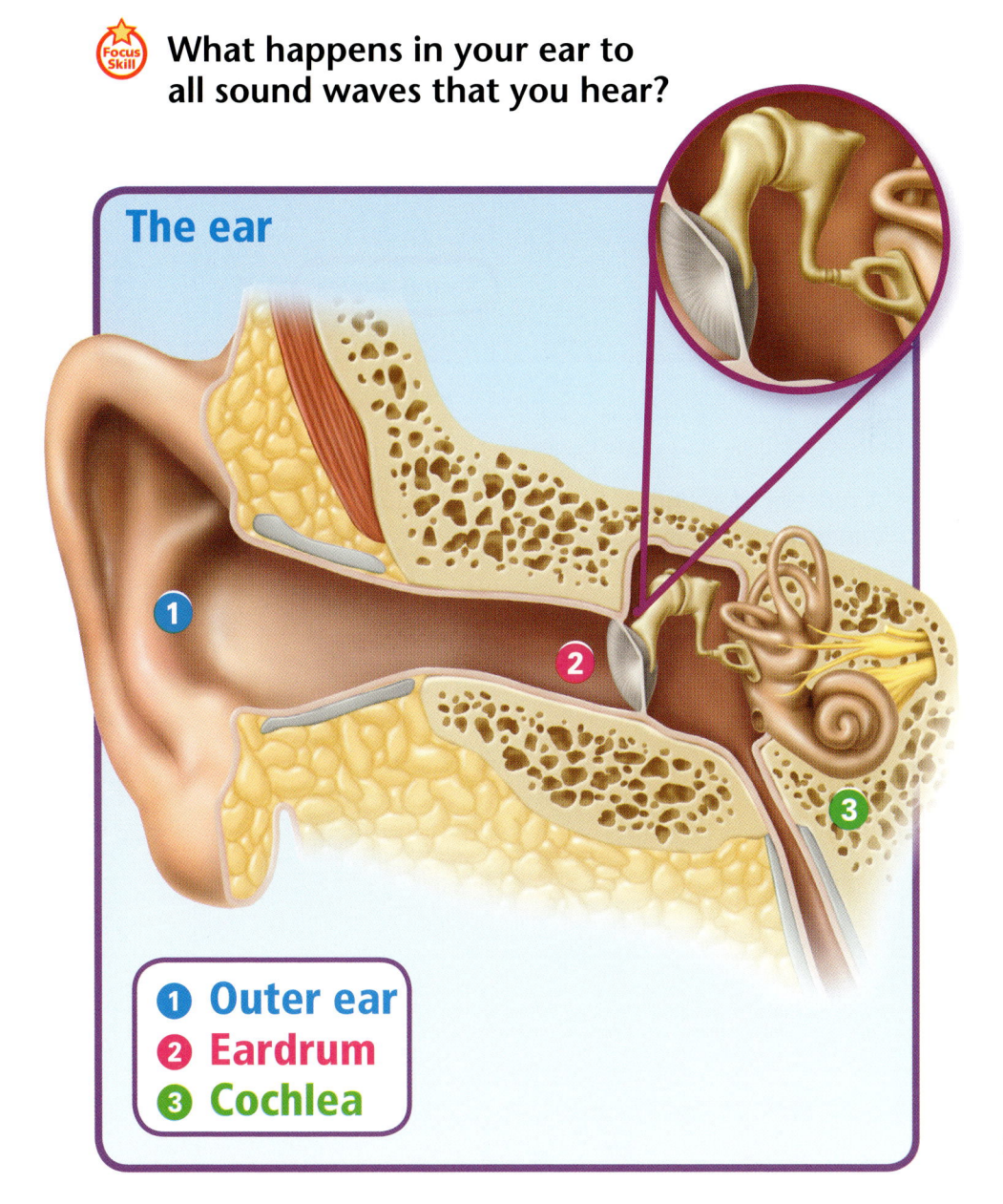

Focus Skill **What happens in your ear to all sound waves that you hear?**

The ear

❶ Outer ear
❷ Eardrum
❸ Cochlea

Reflection

What happens when sound waves strike a surface other than your ear? It depends on the surface. Sound waves bounce off surfaces. This bouncing off of sound waves is called a **reflection**.

If the surface is smooth and flat, sound waves bounce back in the same pattern. You hear the sound again. That is what an echo is.

A pattern of sound waves from a flat, smooth surface ▶

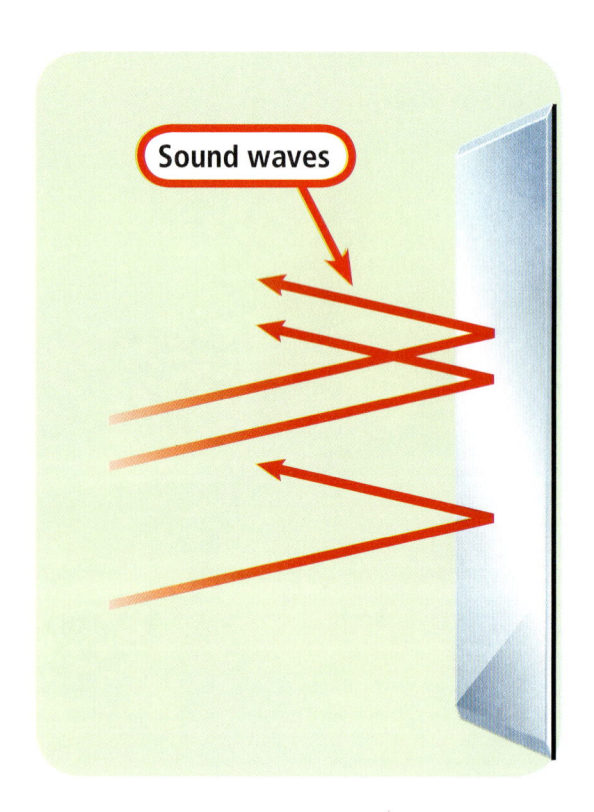

When sound waves strike a rough, uneven surface, the pattern is not kept. Sound waves are scattered all over.

Tell what is different about the two ways sound waves can reflect off a surface.

No pattern of sound waves from a rough, uneven surface ▶

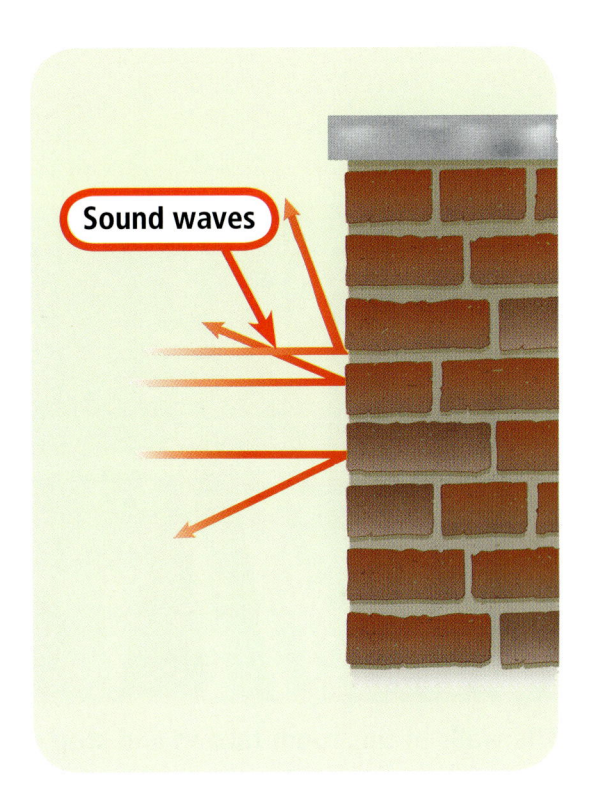

Sound waves

Absorption

Have you ever made sounds in an empty room? It is quite noisy. Sound waves bounce off the walls. If you make the same sounds in a room with carpets and curtains, it is quieter. The carpets and curtains take in the sound. They stop sound waves from reflecting or traveling further. The sound dies out. This process is called **absorption**.

Focus Skill **Tell how absorption and reflection are different.**

The walls of this room take in and stop sound waves.

Transmission

Transmission is the passing of sound waves through matter. Any matter that vibrates can transmit sound. You hear someone talking in another room. Sound waves are transmitted through air in the other room. Then the waves are transmitted through the wall. They travel through air in your room and to your ear.

Focus Skill Tell how transmission and absorption are different.

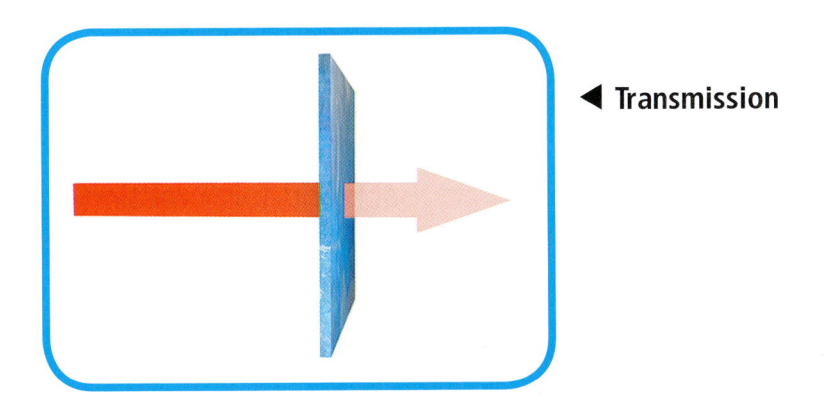

◀ Transmission

Focus Skill Complete these **compare and contrast** statements.

1. All sound waves that you hear make your _____ vibrate.

2. Sound waves that bounce off a smooth wall or a rough wall are both examples of _____.

3. _____ is the stopping of sound waves, but _____ is the passing of sound waves through matter.

GLOSSARY

absorption (ab•ZAWRP•shuhn) The taking in of sound energy by an object

amplitude (AM•pluh•tood) A measure of the amount of energy in a wave

frequency (FREE•kwuhn•see) A measure of the number of waves that pass in a second

intensity (in•TEN•suh•tee) A measure of how loud or soft a sound is

pitch (PICH) A measure of how high or low a sound is

reflection (rih•FLEK•shuhn) The bouncing of light, sound, or heat off an object

transmission (tranz•MISH•uhn) The passing of sound waves through a material

vibration (vy•BRAY•shuhn) A quick back-and-forth motion

wavelength (WAYV•length) The distance between a point on one wave and the identical point on the next wave